BAOFENG RADIO
POCKET GUIDE

Easy setup and operation for new users

Master Your Radio in Minutes

Copyright © 2024 by Jake T Spencer.

All rights reserved. No part of this book may be reproduced or transmitted in any form or by any means, electronic or mechanical, including photocopying, recording, or by any information storage and retrieval system, without permission in writing from the publisher

FIRST EDITION OCT. 2024

Disclaimer:

The content in this book is intended for informational purposes only. The author and publisher are not responsible for any misuse of the information provided. Always follow local laws and regulations related to radio usage.

TABLE OF CONTENTS

1. WELCOME TO THE WORLD OF BAOFENG RADIOS 6

Why Baofeng? (And Why It's Perfect for You) 8

Who Can Use a Baofeng Radio? 9

2. WHAT YOU NEED TO KNOW ABOUT BAOFENG RADIOS 13

The Popular Models (Don't worry, I'll break them down for you!) 13

Key Features Explained Simply 15

Dual-Band Basics: What UHF and VHF Mean for You 18

3. HOW TO PICK THE RIGHT BAOFENG FOR YOU 20

The Best Baofeng for Beginners (No tech skills required!) 20

Options for HAM Radio Enthusiasts 22

If You're an Outdoorsy Type, Here's What You'll Love 24

4: LETS' GET YOUR BAOFENG SET UP 27

Step-by-Step: How to Program Your Radio 27

Using CHIRP Software (It's easier than you think) 29

Frequencies You Can Use Legally (I'll keep you on the right side of the law) 30

Top Emergency Frequencies to Know 32

5: HOW TO USE YOUR BAOFENG IN REAL LIFE 34

Powering On and Setting Up Your Radio 34

Scanning Channels and Sending Your First Message 36

What Affects Your Radio's Range? (And how to get the best signal) 38

Simple Communication Tips for Any Situation 40

A Simple Guide to Baofeng Radio Buttons and Their Functions 44

How to Program Channels on the BaoFeng UV-5R: Using DCS and CTCSS Codes 53

6: BOOSTING YOUR BAOFENG: ACCESSORIES THAT MAKE A DIFFERENCE 57

The Best Antennas for Better Range 58

Extra Batteries and Earpieces You Should Consider 60

Easy Upgrades to Customize Your Radio 63

7: IMPORTANT SAFETY AND LEGAL INFO.................................67

Do You Need a License? (Let's Clear This Up!) 67

Using Your Radio in Emergencies: What You Should Know 70

Staying Safe and Responsible While Using Your Baofeng .72

8: TOP BAOFENG MODELS FOR 2024 76

Baofeng UV-5R: The Budget-Friendly Choice 76

Baofeng UV-82: More Features, Still Simple 78

Baofeng BF-F8HP: Power and Range for Advanced Users 79

Easy Comparison: Which One is Right for You? 81

9: TAKING CARE OF YOUR BAOFENG 83

Cleaning and Storing Your Radio (It's easier than you think) 83

Troubleshooting Common Issues (Don't panic, I'll Walk you through it) 85

Keeping Your Baofeng in Top Shape 87

10: FINAL THOUGHTS 91

Why Baofeng Radios Are a Smart Investment 91

Last-Minute Tips for Making the Most of Your Radio 94

1. WELCOME TO THE WORLD OF BAOFENG RADIOS

A close call in the wilderness serves as a powerful reminder of the importance of reliable communication tools. My colleague Mike, an enthusiastic hiker, found himself lost in a dense forest one weekend. Despite carrying a Baofeng UV-5R radio—a gift he had received months earlier—he had never taken the time

to learn how to operate it. To him, it was just another gadget rather than an essential tool for safety.

As dusk approached, panic set in. Alone and disoriented, he remembered the radio but felt unprepared to use it. In a moment of desperation, Mike decided to give it a try. He turned it on, fumbled with the buttons, and called out for help. To his surprise, an experienced ham radio operator responded, guiding him through how to communicate his location. Thanks to that radio, Mike was rescued by a search team that night.

Mike's experience highlights a crucial lesson: owning a Baofeng radio is only part of the equation; knowing how to use it effectively can be a lifesaver. In this book, you'll learn everything from the basics of selecting the right Baofeng model to step-by-step instructions on programming and operating your radio. We'll explore essential features, accessories that enhance your experience, and crucial safety information. By the end, you'll be empowered to navigate the unexpected confidently and make the most of your Baofeng radio in any situation.

Baofeng radios have gained a reputation for being versatile, affordable, and highly functional, making them a go-to choice for many. But why exactly should you consider

Baofeng, and who are these radios perfect for? Let's make it easier to understand.

Why Baofeng? (And Why It's Perfect for You)

Baofeng radios stand out for a variety of reasons, particularly their **affordability, flexibility, and ease of use**. Whether you're a complete beginner or a seasoned radio enthusiast, Baofeng offers something that fits your needs. Let's take a closer look at why Baofeng is such a solid choice:

1. **Affordability**: One of the main reasons people gravitate toward Baofeng is the price. High-quality radios can be expensive, but Baofeng strikes a balance by providing powerful, reliable radios at a fraction of the cost of other brands. Whether you're on a tight budget or just don't want to overspend, Baofeng gives you excellent value for your money.

2. **User-Friendly**: While some radios require a lot of technical knowledge, Baofeng radios are **straightforward and simple** to operate, even for first-timers. The interface is easy to understand, and there are tons of guides (like this one) to help you learn quickly.

3. **Versatility**: With Baofeng, you're not limited to one type of use. You can switch between UHF and VHF frequencies, use them for HAM radio, emergency communications, hiking trips, or even prepping. This versatility makes Baofeng an all-in-one solution for many different types of radio users.

4. **Durability**: Don't let the affordable price tag fool you — Baofeng radios are **built to last**. Whether you're using it in the rain, in rough outdoor conditions, or just for everyday communication, these radios are designed to hold up.

5. **Customization**: Baofeng radios are also known for their ability to be customized with accessories like better antennas, extended batteries, and earpieces. You can truly make the radio your own and optimize it for your specific needs.

6. **Wide Community Support**: There's a large community of Baofeng users, so if you ever have questions, need advice, or run into trouble, you'll find forums, videos, and plenty of people willing to help you out.

Who Can Use a Baofeng Radio?

Baofeng radios are popular across a wide spectrum of users. Whether you're a **beginner** or someone with experience in radio communication, Baofeng has features that make it appealing to all.

1. **Preppers and Emergency Planners**: For people focused on emergency preparedness, Baofeng radios are an essential tool. These radios can operate on emergency frequencies and stay functional when other forms of communication, like cell phones, are down. Being able to rely on your Baofeng during natural disasters or emergencies can be a lifesaver.

2. **Hikers, Campers, and Outdoor Enthusiasts**: If you spend a lot of time in the wilderness, having a reliable communication device is critical. Baofeng radios offer excellent **range** and can connect with local repeaters to extend that range even further, making them perfect for staying in touch with others during outdoor activities. Plus, their durability makes them suitable for rugged conditions.

3. **HAM Radio Operators**: For those involved in the HAM radio community, Baofeng radios are a popular choice. They provide access to both UHF and VHF frequencies, making them versatile tools for **experimentation and communication**. Many HAM radio operators start their journey with a Baofeng because of its ease of use and affordability.

4. **Event Organizers**: If you're organizing an event where **instant communication** is required, Baofeng radios come in handy. Whether it's a festival, a sports event, or even a large group activity, you can rely on these radios to keep teams connected and coordinated in real-time.

5. **Security Personnel**: Security professionals, from bouncers to neighborhood watch teams, use Baofeng radios for their **clear communication** and ability to cover long distances. It's a reliable way to keep team members informed and respond quickly when needed.

6. **Businesses**: For small businesses looking for affordable communication devices, Baofeng radios are an efficient solution. Whether it's **coordinating staff** across large areas or maintaining contact with delivery drivers, Baofeng radios can help ensure smooth operations.

7. **Families**: If you're taking a family trip, Baofeng radios can be an easy way to stay connected, especially in areas where phone signals are weak. Whether you're hiking together or exploring new places, these radios keep everyone in touch without the need for cell service.

Baofeng radios are not just for the pros — they're designed for anyone who needs reliable communication. Whether you're preparing for emergencies, enjoying the great outdoors, or just looking for a way to stay connected in remote areas, Baofeng radios offer a **simple, powerful, and affordable solution**. And in this guide, you'll learn everything you need to make the most of your Baofeng radio, from choosing the right model to using it effectively in any situation.

2. WHAT YOU NEED TO KNOW ABOUT BAOFENG RADIOS

Now that we've covered why Baofeng radios are a great choice for beginners and casual users, let's dive a little deeper into what makes these radios tick. In this section, we'll take a look at the most popular Baofeng models, explain their key features in simple terms, and break down the basics of dual-band technology, which is actually a major selling point for these devices.

The Popular Models (Don't worry, I'll break them down for you!)

Baofeng radios have a variety of models available, each with its own set of features tailored to different needs. Whether you're a total beginner or a more experienced user, Baofeng has something for you. Let's take a look at some of the most popular models and what makes them unique:

1. **Baofeng UV-5R**: This is probably the most well-known model and is often the go-to choice for beginners. The **UV-5R** is affordable, simple to use, and has all the basic features you'll need in a two-way radio. It's a **dual-band radio** (more on that later), which means it can operate

on two different frequency ranges, making it versatile enough for most common uses. It also has programmable channels, which allows you to store and switch between frequencies with  ease. The UV-5R is great for anyone just getting started with radios, and it's often the first model people recommend because of its combination of price and performance.

2. **Baofeng UV-82**: The **UV-82** is a step up from the UV-5R. It's built with more durability and has a slightly larger body, making it a bit easier to handle, especially for users who need a radio in rugged conditions. This model also comes with a more powerful **transmitter** and a dual push-to-talk (PTT) button, which means you can transmit on two frequencies without having to switch channels manually. This is particularly useful for emergency

responders or anyone coordinating across different groups. If you need something a bit more robust but still easy to use, the UV-82 might be your best bet.

3. **Baofeng BF-F8HP**: For those looking for more power and range, the **BF-F8HP** is a fantastic option. It's similar to the UV-5R but has an upgraded **8-watt transmitter**, which gives you better range and signal strength compared to the standard 4- or 5-watt radios. This model is ideal for users who need to communicate over longer distances or in challenging environments where a stronger signal is necessary. It's also compatible with a wide range of accessories, such as higher-gain antennas, which can further improve your communication capabilities.

Each of these models comes with features that make them suitable for different use cases, so whether you're hiking, prepping for emergencies, or just looking for a reliable radio, there's a Baofeng model that will meet your needs.

Key Features Explained Simply

Baofeng radios come with a variety of features that might sound a bit technical, but don't worry—I'll break them down in simple terms so you can understand how they benefit you.

1. **Dual-Band Capability**: This is one of the standout features of Baofeng radios. Dual-band means the radio can operate on two different frequency ranges: UHF (Ultra High Frequency) and VHF (Very High Frequency). Why does this matter? It allows you to communicate more effectively in different environments. UHF is great for urban areas or places with lots of obstacles (like buildings or trees), while VHF is better for open spaces, such as rural areas or mountains. Having both options available means your Baofeng radio can adapt to various terrains and communication needs.

2. **Programmable Channels**: Baofeng radios come with programmable channels, which allows you to store frequencies for quick access. You can program the radio manually using the keypad or use free software like **CHIRP** to program your radio via a computer. This is particularly useful if you have multiple frequencies you use regularly. Instead of manually tuning into a frequency each time, you can just switch between channels with the push of a button.

3. **Emergency Features**: Many Baofeng radios come with built-in emergency features like a **flashlight** and an **emergency alert system**. The flashlight can be handy if you're using the radio in low-light conditions,

and the emergency alert system can be used to send a distress signal if you find yourself in a dangerous situation. This is a useful feature for anyone who spends a lot of time outdoors or in environments where you might need help quickly.

4. **Long Battery Life**: Baofeng radios typically come with **rechargeable lithium-ion batteries** that last for several hours of continuous use. Some models also offer extended battery options for even longer usage. This is especially useful if you plan to use the radio for extended periods, such as during a hike or a long outdoor adventure. Additionally, many Baofeng radios allow you to carry **spare batteries**, so you don't have to worry about running out of power when you need it the most.

5. **Compact and Lightweight**: One of the reasons Baofeng radios are so popular is because they are both **compact** and **lightweight**. This makes them easy to carry around, whether you're putting them in a backpack or clipping them onto your belt. Despite their small size, these radios pack a punch in terms of features, making them a great option for anyone who values portability.

6. **Customizable Accessories**: Baofeng radios can be upgraded with a variety of **accessories** to improve performance. For example, you can swap out the standard antenna for a longer one that increases your range or add an external microphone for hands-free use. There are also plenty of battery and earpiece options that can enhance your radio experience. This flexibility makes Baofeng radios not only affordable but also customizable to suit your specific needs.

Dual-Band Basics: What UHF and VHF Mean for You

A key feature of Baofeng radios is their **dual-band capability**, meaning they can operate on both **UHF** (Ultra High Frequency) and **VHF** (Very High Frequency) bands. If you're new to radios, these terms might sound a bit confusing, but I'll explain them simply and show you why it matters for your day-to-day use.

> **UHF (Ultra High Frequency)**: UHF ranges from 400 MHz to 520 MHz. This frequency band works best in **urban environments** or areas with lots of **obstacles**, like buildings, trees, or hills. UHF signals have shorter wavelengths, which means they can penetrate through walls and other obstructions more easily. If you're using

your Baofeng radio in a city or around large structures, UHF will give you a clearer signal.

- **VHF (Very High Frequency)**: VHF ranges from 136 MHz to 174 MHz. This frequency band is ideal for **open spaces** like **rural areas** or **mountains**. VHF signals travel further in areas without many obstructions, so if you're out in the wilderness, a VHF frequency will help you communicate over long distances. However, VHF isn't as good at penetrating through dense obstacles like buildings, so it's better suited for outdoor use.

Why is this important? With a dual-band Baofeng radio, you can choose which band to use depending on your surroundings. If you're in an urban area, you can switch to UHF for better performance, and if you're out in the countryside or on a hiking trip, you can use VHF for longer-range communication. This versatility means you're always prepared, no matter where you are.

In conclusion, knowing the popular Baofeng models, understanding their key features, and grasping the basics of dual-band technology will help you make the most out of your Baofeng radio. Whether you're just getting started or looking to upgrade, Baofeng offers a range of options and features that can meet your communication needs without overwhelming you with complexity.

3. HOW TO PICK THE RIGHT BAOFENG FOR YOU

With so many models available, it can be overwhelming to figure out which Baofeng radio is right for you. Don't worry—I've got you covered! Whether you're a complete beginner, an experienced HAM radio enthusiast, or someone who enjoys spending time in the great outdoors, there's a Baofeng radio that's perfect for your needs. In this section, I'll Walk you through the best options for each type of user so you can make the right choice without any confusion.

The Best Baofeng for Beginners (No tech skills required!)

If you're new to the world of two-way radios and don't have any technical experience, you'll want a Baofeng model that's simple to use yet still powerful enough to get the job done. The best Baofeng radio for beginners is the **Baofeng UV-5R**.

Here's why:

1. **Easy to Use**: The UV-5R is straightforward, with a simple button layout that's easy to understand. You don't need to be a tech expert to figure out how to turn

it on, switch channels, or communicate. The manual is helpful, and there are tons of online tutorials if you need extra help.

2. **Affordable**: One of the biggest draws of the UV-5R is its price. It's one of the most affordable dual-band radios on the market, making it a great option if you're just starting out and don't want to invest a lot of money upfront.

3. **Dual-Band Capabilities**: Even though the UV-5R is designed for beginners, it still has powerful features like **dual-band capability** (UHF and VHF). This means you can communicate across different frequencies, whether you're in an urban area or out in the open countryside.

4. **Programmable Channels**: The UV-5R comes with **128 programmable channels**, so you can easily store and access your favorite frequencies. Whether you're chatting with friends, coordinating in an emergency, or listening in on public frequencies, the UV-5R makes it easy.

If you're looking for a reliable, easy-to-use radio that doesn't require any technical know-how, the **UV-5R** is the perfect choice. It's a solid introduction to two-way radios and offers all the essential features without being overwhelming.

Options for HAM Radio Enthusiasts

For those who have a bit more experience and are licensed HAM radio operators, Baofeng offers more advanced models with extra features and capabilities. Two of the best options for HAM radio enthusiasts are the **Baofeng UV-82** and the **Baofeng BF-F8HP**.

1. **Baofeng UV-82**:

    - **Dual PTT Button**: One unique feature of the UV-82 is its **dual push-to-talk (PTT) button**, which allows you to transmit on two different frequencies without having to switch channels manually. This is especially useful for HAM operators who need to communicate on multiple frequencies at once.

    - **Durability**: The UV-82 is built to be a bit more rugged than the UV-5R, so it's better suited for users who need a radio that can withstand rough conditions. Whether you're using it in an emergency situation or during outdoor adventures, the UV-82 is designed to handle it.

    - **Strong Transmitter**: The UV-82 has a stronger transmitter than the UV-5R, which means you'll get a better signal and longer range, perfect for HAM radio enthusiasts who need reliable communication over greater distances.

2. **Baofeng BF-F8HP**:

   ➢ **More Power**: The **BF-F8HP** offers a significant upgrade in terms of power, with an **8-watt transmitter** (compared to the 5-watt transmitter of the UV-5R). This extra power provides greater range and clearer signals, especially in areas where communication might otherwise be difficult.

   ➢ **Customizable Settings**: The BF-F8HP allows you to adjust settings for even more control over how your radio operates. You can tweak the power settings, frequency ranges, and more, making it an excellent choice for HAM operators who want to fine-tune their radio experience.

   ➢ **Extended Battery Life**: The BF-F8HP comes with a larger **2000mAh battery**, providing extended usage time. For long conversations or extended trips, this extra battery life can make all the difference.

For HAM radio enthusiasts who want more advanced features, the **UV-82** and **BF-F8HP** are excellent choices.

They offer greater range, more power, and the flexibility to handle a variety of communication needs.

If You're an Outdoorsy Type, Here's What You'll Love

For those who love outdoor activities like hiking, camping, or off-roading, having a reliable communication device is essential. You need a radio that can handle rugged conditions and provide strong signals in remote areas. Fortunately, Baofeng has radios that are perfect for outdoor enthusiasts.

1. **Baofeng UV-82**:

    - **Rugged Build**: As mentioned earlier, the UV-82 is more durable than the UV-5R, which makes it a great choice for outdoor use. If you're hiking, camping, or engaging in any activity where your radio might get knocked around, the UV-82 can handle the abuse.

    - **Dual PTT for Group Communication**: If you're hiking or camping with a group, the **dual PTT button** on the UV-82 is a handy feature. You can stay in touch with different members of your group on separate frequencies without needing to change channels manually.

2. **Baofeng BF-F8HP**:

- **Extended Range for Remote Areas**: With its **8-watt transmitter**, the **BF-F8HP** offers extended range, making it a great option if you're hiking in remote areas or places with rough terrain. The extra power helps ensure that your signal reaches farther, so you can stay in touch with others even when you're deep in the wilderness.

- **Long Battery Life for Extended Adventures**: Outdoor activities often mean you'll be away from power sources for long periods, so the **extended battery life** on the BF-F8HP is a big plus. The larger battery means you can use your radio for longer without needing to recharge, perfect for multi-day trips.

3. **Accessory Compatibility**: Both the UV-82 and BF-F8HP are compatible with a range of accessories, such as **high-gain antennas** that increase your communication range and **hands-free earpieces** that let you use your radio while on the move. These accessories are especially useful for outdoor activities where you might need extra range or hands-free convenience.

For outdoor enthusiasts, the **UV-82** and **BF-F8HP** provide ruggedness, extended range, and long-lasting battery life, making them excellent choices for staying connected in the great outdoors.

Picking the right Baofeng radio doesn't have to be complicated. Whether you're a beginner, a HAM radio enthusiast, or someone who loves spending time outdoors, there's a Baofeng model that's perfect for you. The **UV-5R** is a great option for beginners, offering ease of use and affordability. If you need something more advanced, the **UV-82** and **BF-F8HP** provide extra power, durability, and features tailored to more experienced users. With Baofeng, you can find a reliable, affordable radio that fits your specific needs.

4: LETS' GET YOUR BAOFENG SET UP

Now that you've chosen the right Baofeng radio for your needs, it's time to set it up. Don't worry—it's much easier than it sounds, and I'll Walk you through every step. By the end of this chapter, you'll know how to program your radio, use CHIRP software, and make sure you're using legal frequencies. I'll also share some of the top emergency frequencies you should know. Let's go!

Step-by-Step: How to Program Your Radio

The first thing you need to do when you get your Baofeng radio is to **program it**. Programming your radio means setting up the channels and frequencies you'll use. It might seem a little complicated at first, but I'll break it down into simple steps so anyone can do it.

1. **Turn on the Radio**: Hold down the power button to turn on your Baofeng radio. You should see the screen light up, and the default frequency will appear.

2. **Switch to VFO Mode**: You'll need to be in **VFO (Variable Frequency Oscillator) mode** to manually input frequencies. To switch modes, press the **VFO/MR button** on your radio. The screen will display frequencies instead of channel numbers when you're in VFO mode.

3. **Enter a Frequency**: Use the keypad on your radio to enter the frequency you want to program. For example, if you're entering 146.520 MHz (a common emergency frequency), you would type in the numbers directly.

4. **Save the Frequency to a Channel**: Once you've entered the frequency, you'll need to save it to a channel for easy access later. To do this:

   ✓ Press the **Menu button**.

   ✓ Scroll through the options using the arrow keys until you find **Menu 27 (MEM-CH)**, which is the option for saving channels.

   ✓ Press **Menu** again, and then select the channel number you want to assign the frequency to (e.g., Channel 1).

   ✓ Press **Menu** once more to confirm, then exit the menu.

5. **Repeat for Other Frequencies**: You'll need to repeat this process for any additional frequencies you want to save. Don't worry—it gets faster once you've done it a few times.

And there you have it! Your Baofeng radio is now programmed with the frequencies you need. If you ever want to change them, just follow the same steps to reprogram the channels.

Using CHIRP Software (It's easier than you think)

If manually entering frequencies feels too tedious, don't worry! There's an easier way to program your Baofeng radio: **CHIRP software**. This free software allows you to program your radio using your computer, which can save a lot of time.

Here's how to use CHIRP:

1. **Download CHIRP**: Head over to the official CHIRP website and download the software that matches your operating system (Windows, Mac, or Linux).

2. **Connect Your Baofeng to Your Computer**: You'll need a **programming cable** to connect your radio to your computer. Make sure the cable is compatible with your Baofeng model (most Baofeng programming cables work across multiple models). Plug one end into your radio's speaker/mic jack and the other into a USB port on your computer.

3. **Install Drivers**: If this is your first time using the programming cable, you may need to install the

appropriate drivers. CHIRP will guide you through this process, and the drivers should come with the cable.

4. **Download Your Radio's Settings**: Once everything is connected, open CHIRP and click **Download from Radio**. Follow the prompts to select your radio model and confirm the download.

5. **Program Your Frequencies**: Now that your radio's settings are displayed in CHIRP, you can easily enter new frequencies into the software. Simply type in the frequencies you want for each channel, and CHIRP will handle the rest.

6. **Upload the Settings to Your Radio**: After you've entered all your frequencies, click **Upload to Radio** to send the settings back to your Baofeng. Your radio will now be fully programmed, and you're ready to go!

Using CHIRP makes it easy to program multiple frequencies quickly, and it's especially helpful if you plan to make regular changes to your radio's channels.

Frequencies You Can Use Legally (I'll keep you on the right side of the law)

One of the most important things to know when using a Baofeng radio is that not all frequencies are legal for everyone to use. Some frequencies are restricted to licensed operators, such as HAM (Amateur Radio) operators, while others are available for public use.

Here's a quick breakdown of the frequencies you can legally use without a HAM license:

1. **FRS (Family Radio Service)**: These frequencies are legal for anyone to use without a license. They're often used for short-range communication between family members or in small groups, and they're perfect for things like hiking, camping, or neighborhood coordination.

    ➤ Example: 462.5625 MHz (Channel 1 on most FRS radios)

2. **GMRS (General Mobile Radio Service)**: GMRS frequencies require a license to use, but they offer greater range than FRS. You can apply for a GMRS license through the FCC, and it's valid for your entire family.

    ➤ Example: 462.550 MHz (Channel 15 on most GMRS radios)

3. **MURS (Multi-Use Radio Service)**: MURS is another unlicensed frequency band that can be used for longer-range communication. These frequencies are commonly used for business or personal communication over longer distances.

> Example: 151.820 MHz (MURS Channel 1)

4. **HAM Frequencies**: If you have a HAM license, you'll have access to a much wider range of frequencies. HAM operators can communicate over longer distances and use more advanced radio equipment. If you're interested in getting your HAM license, it's a great way to expand your radio skills.

Always make sure to double-check the regulations in your area and ensure that you're using frequencies legally. It's important to stay on the right side of the law when using two-way radios.

Top Emergency Frequencies to Know

Having access to emergency frequencies is one of the most valuable aspects of owning a Baofeng radio. In a disaster situation, cell phones may not work, but your Baofeng can keep you connected to essential services. Here are some of the top emergency frequencies you should program into your radio:

1. **146.520 MHz**: This is the national calling frequency for HAM radio operators in the U.S. It's monitored by many emergency services, so it's a good frequency to have programmed in case of emergencies.

2. **462.675 MHz**: This is the **National Emergency Channel** for GMRS radios. If you have a GMRS license, this frequency can connect you with emergency services or other GMRS operators during a disaster.

3. **121.500 MHz**: Known as the **International Aeronautical Emergency Frequency**, this is monitored by aircraft and air traffic control and can be used in life-threatening situations.

4. **156.800 MHz**: This is the **Marine VHF Channel 16**, which is the international distress frequency for mariners. If you're on a boat or near the coast, this is a crucial frequency to have.

By programming these emergency frequencies into your Baofeng radio, you'll be prepared to communicate with the right people in case of an emergency, whether you're in the wilderness or in an urban environment during a disaster.

Getting your Baofeng radio set up doesn't have to be a difficult process. Whether you're programming it manually or using CHIRP software, following these simple steps will ensure your radio is ready to go. Make sure you're using the right frequencies to stay legal, and don't forget to program some emergency channels—you never know when they might come in handy!

5: HOW TO USE YOUR BAOFENG IN REAL LIFE

Once your Baofeng radio is all set up, the real fun begins—using it in everyday situations. Whether you're a total beginner or have some experience with radios, I'll walk you through how to power on your radio, scan channels, send messages, and improve your range. By the end of this chapter, you'll feel confident communicating in a variety of scenarios. So let's get started!

Powering On and Setting Up Your Radio

The very first thing you'll need to do is power on your Baofeng radio. This step is simple, but it's always good to go over the basics so that you know you're getting everything right.

34 *Baofeng Radio Pocket Guide*

1. **Turning on the Radio**: To power up your Baofeng, **turn the Power/Volume knob** (located on the top of the device) clockwise. You'll hear a click, and a voice will announce that the radio is on. The screen will light up, showing the default frequency or channel number.

2. **Battery Check**: Before you do anything else, make sure your battery is charged. If you're running low on battery, it's better to swap in a fresh one now rather than in the middle of using the radio. Baofeng radios usually come with rechargeable batteries, so you can just plug your radio into a charger when needed.

3. **Check the Antenna**: For optimal performance, ensure your antenna is securely attached. If it's loose, your signal strength will suffer, and you might not be able to reach other radios as effectively.

4. **Set the Volume**: Adjust the volume to a comfortable level using the volume knob, which is typically located on the same button as the power switch. Make sure it's loud enough to hear, but not so loud that it causes distortion.

5. **Verify Frequency or Channel**: Take a look at the screen to make sure you're on the right frequency or

channel. You'll likely be using either **VFO mode** (where you manually enter frequencies) or **MR mode** (where you're using pre-programmed channels). You can switch between these modes by pressing the **VFO/MR button** on the front of the radio.

Once these basic checks are done, your Baofeng is ready to use!

Scanning Channels and Sending Your First Message

Now that your radio is on and set up, the next step is learning how to communicate. One of the key features of any two-way radio is the ability to scan channels and send out messages. With your Baofeng, this process is both simple and efficient.

1. **Scanning Channels**:

    ➤ To scan for active channels, press the **SCAN button** (often labeled as **A/B** on the UV-5R model). This will automatically search through all available frequencies and stop when it detects an active signal.

    ➤ You can also manually input the frequency you want to listen to or communicate on using the keypad. For example, if you want to tune into the **emergency** frequency **146.520 MHz**, simply enter the numbers and the radio will switch to that frequency.

- While scanning, keep an ear out for active conversations. Once the radio stops on a frequency where someone is talking, it will pause for you to listen in.

2. **Sending Your First Message**:

- Find a frequency or channel that isn't currently in use. It's a good habit to listen for a moment before transmitting, as you don't want to accidentally interrupt someone's conversation.

- Once you've confirmed the channel is clear, hold down the **PTT (Push-To-Talk) button** on the side of your radio. While holding the button, speak clearly and directly into the radio's microphone.

- Keep your message short and to the point. For example, if you're reaching out to a friend, you could say something like: "This is [Your Name], does anyone copy?"

- Release the **PTT button** when you're finished speaking, and wait for a response.

That's it! You've now successfully scanned channels and sent your first message. It's a straightforward process, but one that forms the backbone of using your Baofeng in everyday communication.

What Affects Your Radio's Range? (And how to get the best signal)

One of the most frequently asked questions about Baofeng radios is: **What affects my radio's range?** Understanding this is crucial, as it will help you get the best possible signal in a variety of environments. While Baofeng radios are powerful for their size and price, a few factors can influence how far your signal will travel.

1. **Antenna Quality**:

    - The type of antenna you're using plays a big role in the range of your Baofeng radio. The standard "rubber duck" antenna that comes with most models is decent, but upgrading to a **longer, high-gain antenna** can significantly increase your range.
    - If you're struggling with range issues, consider swapping out the stock antenna for one like the **Nagoya NA-771**. This can boost your signal, especially in challenging environments.

2. **Line of Sight (LOS)**:

- Radios work best when there's a **clear line of sight** between you and the other radio you're communicating with. That means the fewer obstacles between you (like buildings, trees, or hills), the farther your signal can travel.

- If you're in a wide-open area, like a field or on top of a hill, you'll likely get much better range than if you're in a dense urban area or surrounded by heavy foliage. In some cases, even positioning yourself a few feet higher (like on top of a building or hill) can make a big difference.

3. **Weather Conditions**:

    - Believe it or not, the weather can also impact your radio's performance. **Rain, fog, and other moisture** in the air can weaken your signal, especially over long distances. While Baofeng radios can still function in bad weather, you might notice a reduction in range during heavy storms or in humid conditions.

4. **Battery Life**:

    - A radio with a **full battery** will typically transmit better than one that's running low on power. If you're noticing reduced range, check your battery level and consider charging or swapping out the battery. Having a **spare battery** is always a good idea, especially if you're out in the field for an extended period of time.

5. **Interference**:

   ➤ Radio waves can sometimes encounter **interference** from other electronic devices, especially in urban areas. This interference can reduce your range and make communication difficult. If you're experiencing interference, try moving to a different location or switching to a different frequency.

   ➤ If you're in a crowded area with lots of radio traffic, you might need to adjust your settings or find a less congested channel to improve your range.

By understanding these factors, you'll be able to get the most out of your Baofeng radio's range. Experiment with different environments and settings to see what works best for you.

Simple Communication Tips for Any Situation

Whether you're using your Baofeng for emergency situations, casual communication, or outdoor adventures, there are a few key tips that will help you communicate more effectively. These tips will not only improve your experience but also ensure you're getting the best use out of your radio.

1. **Speak Clearly and Slowly**:

    ✓ When using any two-way radio, it's important to speak **clearly and at a moderate pace**. If you talk too fast, the person on the other end might not catch everything you're saying. Take your time and make sure your message is easy to understand.

    ✓ Don't hold the radio too close to your mouth—about **2-3 inches away** is ideal. This ensures that your voice comes through clearly without overwhelming the microphone.

2. **Use Short, Simple Messages**:

    ✓ Radio communication is most effective when you keep your messages **short and to the point**. Long-winded conversations can be confusing and may lead to misunderstandings.

    ✓ For example, instead of saying: "Hey, I'm trying to find out if you're available for a quick chat on the radio," try saying: "This is [Your Name], are you available to copy?"

3. **Identify Yourself**:

    ✓ Always start your message by identifying yourself. This helps the person on the other end know who's speaking, especially if multiple people are using the same channel.

- ✓ For example, you can say: "This is [Your Name] calling [Their Name]. Do you copy?"

4. **Listen Before Transmitting**:

- ✓ One of the golden rules of radio communication is to always **listen before you transmit**. You never want to interrupt someone who's already speaking, as this can lead to confusion and frustration.

- ✓ Take a few moments to listen to the channel and make sure it's clear before sending your message. If someone is already talking, wait until they're finished before you press the PTT button.

5. **Acknowledge Messages**:

- ✓ When someone sends you a message, it's good practice to **acknowledge** that you've received it. This confirms to the other person that their message got through and that you're ready to respond.

- ✓ You can simply say: "Copy that," or "Received," to let them know you're on the same page.

6. **Repeat Critical Information**:

- ✓ If you're sharing important information, like an address or a set of coordinates, it's helpful to **repeat it** back to the person to ensure accuracy.

- ✓ For example, if someone gives you an address, repeat it back to them: "Copy that, you're at 123 Main Street, correct?" This helps prevent misunderstandings, especially in time-sensitive situations.

7. **Use Call Signs (If Necessary)**:

   - ✓ In some situations, particularly if you're using HAM radio frequencies, it's common to use **call signs** instead of real names. This is especially important for licensed HAM operators, as it's a regulatory requirement.

   - ✓ If you're new to using call signs, don't worry—just make sure to identify yourself with your assigned call sign when transmitting on HAM frequencies.

8. **Stay Calm in Emergencies**:

   - ✓ If you're using your Baofeng radio in an emergency situation, it's crucial to remain **calm and composed**. Panicking can make it difficult for others to understand your message, and it can lead to mistakes.

- ✓ Take a deep breath, think about what you need to say, and communicate your message as clearly as possible. If you need help, give your location and the nature of the emergency first.

9. **Know When to Switch Frequencies**:

    - ✓ If you're using a crowded frequency, it might be beneficial to **switch to a different one** to reduce interference and improve communication quality. Have a list of backup frequencies programmed into your radio so you can easily switch if needed.

10. **Keep Conversations Professional (When Required)**:

    - ✓ If you're using your Baofeng in a professional setting, such as during an event or for work-related communication, it's important to **keep conversations brief and professional**.
    - ✓ Avoid unnecessary chatter and stick to the information that's directly relevant to the task at hand.

A Simple Guide to Baofeng Radio Buttons and Their Functions

Below is a detailed breakdown of each button on the **Baofeng UV-5R** model, explaining how they work, what

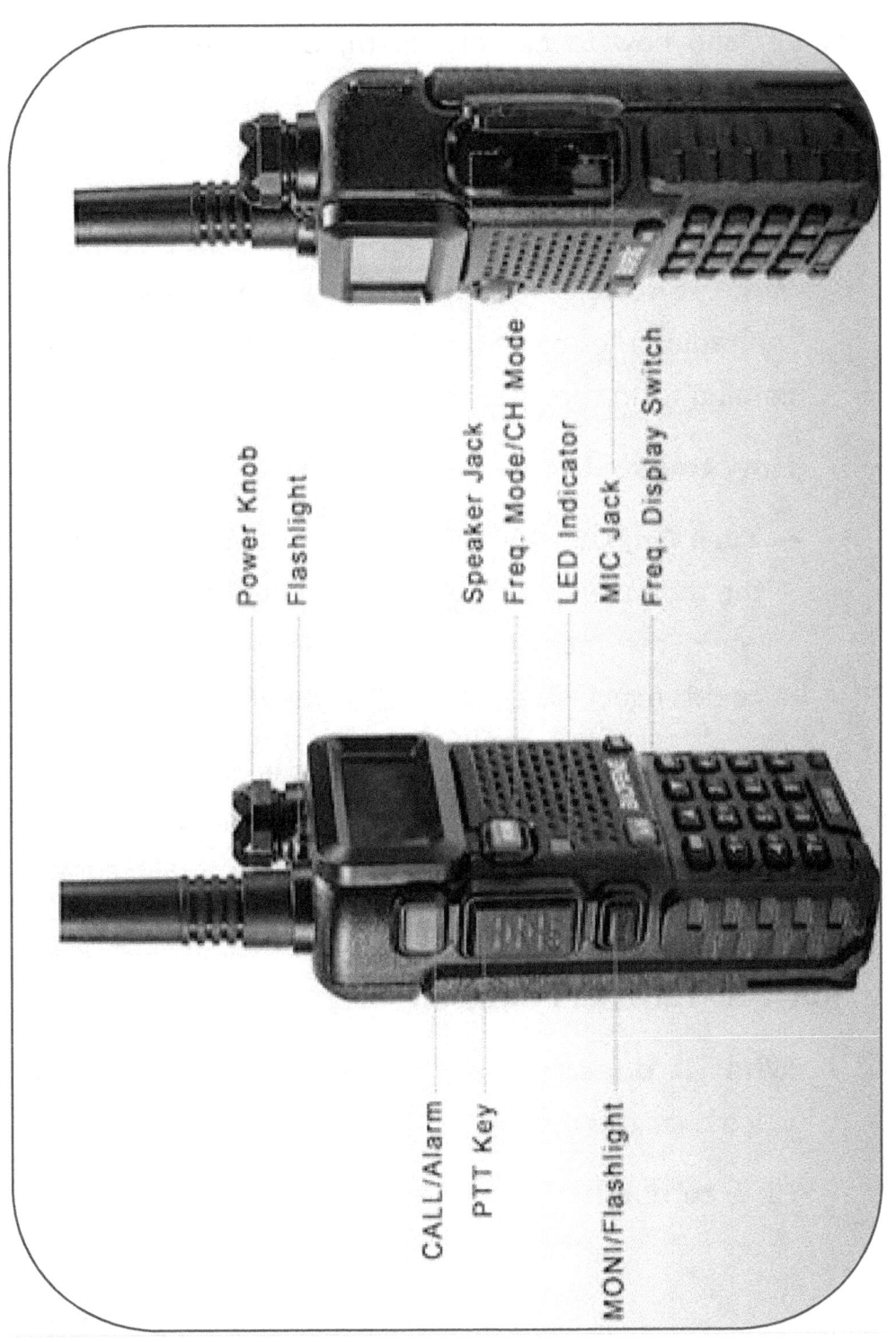

they do, and how to operate them, even if you've never touched a radio before. We'll keep it simple and easy to follow, just like a conversation!

1. Power/Volume Knob

- **What it does**: This knob controls two things: turning the radio on or off and adjusting the volume of sound coming from the speaker.

- **How to use it**:

    - **Turn On/Off**: Turn the knob clockwise to power on the radio. Keep turning until you hear a click – that's when the radio turns on. To turn it off, rotate it counterclockwise until it clicks again.

    - **Adjust Volume**: While the radio is on, keep turning the knob clockwise to increase the volume and counterclockwise to lower it. Start at a low volume, and increase as needed.

2. Push-to-Talk (PTT) Button

- **What it does**: This button lets you transmit your voice over the radio.

- **How to use it**:

> Press and **hold** this button when you want to talk. When you're done talking, release the button. This action switches the radio back to receiving mode so you can hear replies. Think of it as a "talk" button – press to talk, let go to listen.

3. Monitor Button

- **What it does**: This button temporarily disables the squelch, allowing you to listen to weak signals or static.

- **How to use it**:

    > Hold this button to open up the squelch (static filtering) and listen to weaker signals. This is useful if you're trying to catch a faint transmission that's not coming through clearly due to distance or interference.

4. A/B Button

- **What it does**: The screen on your UV-5R has two lines, and this button lets you switch between them.

- **How to use it**:

    > Press the **A/B** button to toggle between the upper and lower lines on the display. Each line can show a different frequency or channel, which is useful for

monitoring two channels at once. The active line will have a little arrow next to it, meaning that's the line you're currently controlling.

> For example, if you want to change the frequency on the upper line, press **A/B** to select it, then enter the frequency using the keypad.

5. VFO/MR Button

- **What it does**: This button toggles between two modes: VFO (Variable Frequency Operation) and MR (Memory Recall).

- **How to use it**:

    > In **VFO Mode**, you can manually enter a frequency. This is helpful when you want to directly input or adjust a specific frequency for communication.

    > In **MR Mode**, you switch to the saved channels in your memory. If you've saved frequencies before, you can easily recall them by cycling through using the up and down arrow keys. Press **VFO/MR** to switch between these two modes.

6. Keypad Buttons

- **What they do**: The keypad is like a number pad on a phone, used for entering frequencies, changing channels, and navigating the menu.

- **How to use them**:

  - **Entering Frequencies**: In VFO mode, you can type in a frequency directly using the number keys. For example, if you want to enter the frequency 146.520 MHz, you'd press **1-4-6-5-2-0**.

  - **Navigating Menus**: Use the number keys to select different options from the radio's menu. When you press the **Menu** button, each function has a number associated with it, which you can select using the keypad.

7. Menu Button

- **What it does**: This button opens up the settings menu, where you can adjust various features like squelch level, power output, and more.

- **How to use it**:

  - Press **Menu** to open the list of settings. Use the up/down arrow keys to scroll through options or enter a specific menu number using the keypad.

➢ After selecting a menu item, press **Menu** again to adjust the setting, then press **Exit** to close the menu.

8. Exit Button

- **What it does**: Exits the current menu or operation.

- **How to use it**:

    ➢ If you've entered a menu or a mode that you want to leave, press the **Exit** button. This takes you back to the main display without saving any unwanted changes.

9. Call Button

- **What it does**: This button sends a call tone or alarm to other radios on the same frequency.

- **How to use it**:

    ➢ Press the **Call** button to send a pre-set tone to alert others on the same frequency. This is useful in emergency situations to grab someone's attention quickly.

10. Side Buttons (Programmable)

- **What they do**: These are additional buttons on the side of the radio that you can program for different tasks like scanning channels or toggling the flashlight.

- **How to use them**:

  > You can assign different functions to these buttons through the menu or software (like CHIRP). For example, you can set one button to activate the radio's built-in flashlight or to start scanning for active frequencies.

Functions of the 2 Display Lines on the Screen

Your UV-5R has two display lines, and here's how they work:

Upper Line (Line 1) and Lower Line (Line 2)

> **What they show**: These lines display the active frequencies or channel names. You can monitor two frequencies or channels

simultaneously, which is one of the best features of Baofeng radios.

- **How to use them:**
  - ❖ The line with the little arrow next to it is the one you're currently controlling. Use the **A/B** button to switch between the two lines. For instance, you might have your local HAM frequency on the top line and an emergency frequency on the bottom. You can quickly toggle between them depending on which one you want to listen to or transmit on.
  - ❖ Each line can show either a frequency or a channel number, depending on whether you're in VFO (frequency mode) or MR (channel mode).

Quick Tip:

To switch the active line for transmitting (the one with the arrow), just press **A/B**. If you want to change the frequency or scan on that line, make sure the arrow is next to it.

Putting It All Together:

By understanding these buttons and how to operate them, you'll be able to confidently set up and use your Baofeng UV-5R. Whether you're programming a frequency, scanning channels, or sending a message, this guide ensures you'll know what each button does and how to use it effectively.

Using your Baofeng radio in real-life situations doesn't have to be complicated. By following these steps for powering on, scanning channels, sending messages, and improving your range, you'll be able to communicate effectively in a variety of settings. Remember to keep your messages clear and concise, always listen before transmitting, and be mindful of your surroundings and signal range.

By mastering these basics, you'll be ready for anything—whether it's a casual chat with friends, coordinating an outdoor adventure, or staying connected in an emergency.

How to Program Channels on the BaoFeng UV-5R: Using DCS and CTCSS Codes

Programming channels on the BaoFeng UV-5R using DCS (Digital Coded Squelch) and CTCSS (Continuous Tone-Coded Squelch System) is a way to set up specific privacy codes, allowing communication on a shared frequency without interference from other users. Here's a simple step-by-step guide for you to get this done:

Step 1: Enter Frequency Mode

- Turn on your Baofeng UV-5R.

- Press the **[VFO/MR]** button to switch the radio into **Frequency (VFO) Mode**.

Step 2: Input the Frequency

- Use the keypad to input the frequency you want to program (for example, 145.600 MHz).

- This is the frequency where you'll transmit and receive.

Step 3: Set CTCSS or DCS Codes

- To set **CTCSS (analog tone)**:

    - Press the **[MENU]** button.

    - Scroll (using the **[UP/DOWN]** arrow keys) to Menu Option 11, which is **T-CTCS** for transmit CTCSS.

    - Press **[MENU]** again to select it.

    - Use the arrow keys to choose your desired CTCSS code (e.g., 67.0 Hz). Press **[MENU]** to confirm.

    - You can also set a **receive CTCSS** code by going to **R-CTCS** in the menu and following the same steps.

- To set **DCS (digital squelch code)**:

- ✓ Press **[MENU]** and scroll to Menu Option 12, **T-DCS** (for transmit).

- ✓ Press **[MENU]** again to select it.

- ✓ Scroll through the available DCS codes using the arrow keys. Select the code you need and press **[MENU]** to confirm.

- ✓ You can set the **receive DCS** code by selecting **R-DCS** in the menu and following the same steps.

Step 4: Save the Frequency to a Channel

- After setting up the frequency, CTCSS, or DCS codes, you'll want to save this to a channel for easy access.

  - ➤ Press **[MENU]** and scroll to Menu Option 27 (**MEM-CH**).

  - ➤ Choose an empty channel number (e.g., 001-127).

  - ➤ Press **[MENU]** to save.

Step 5: Switch to Channel Mode

- Press **[VFO/MR]** to return to Channel Mode, and your newly programmed channel with CTCSS or DCS codes should be available for use.

Test Your Setup

- Test your setup by trying to communicate with someone using the same frequency and CTCSS/DCS codes. If done correctly, only radios using the same codes can hear or transmit on the frequency.

This guide simplifies the process for beginners and ensures you're using the radio in a way that keeps unwanted transmissions from interfering with your communication.

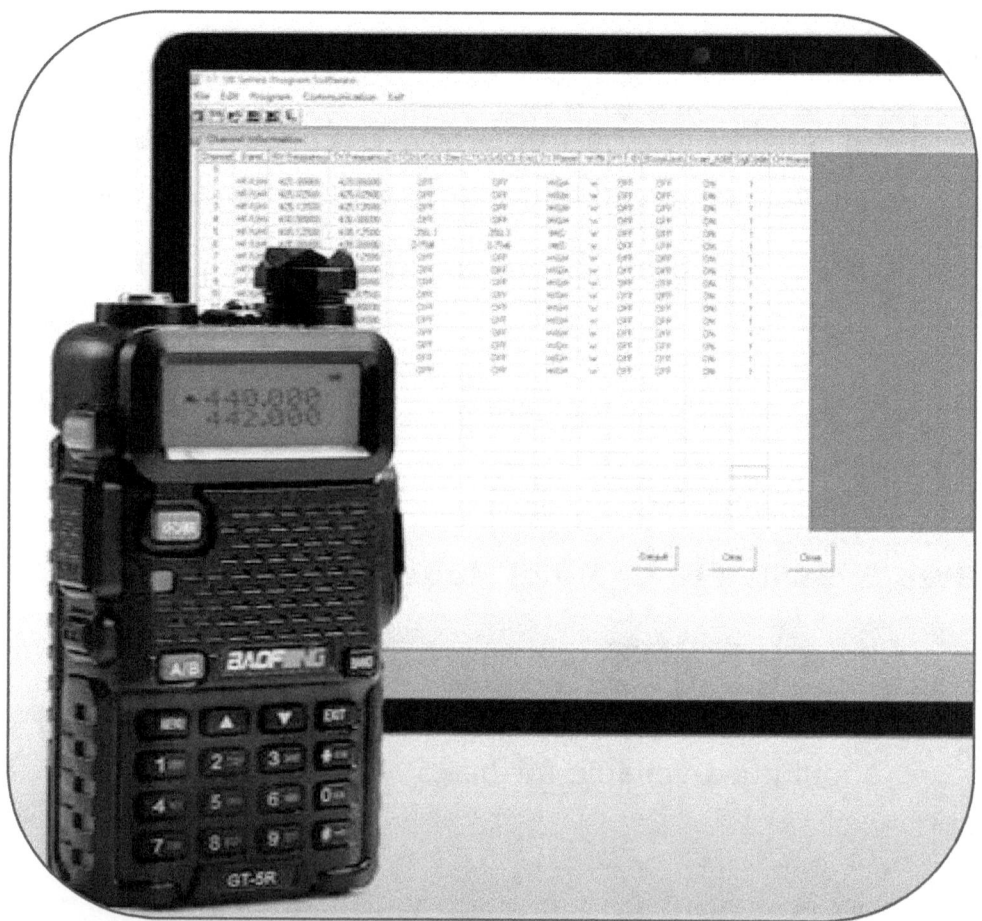

6: BOOSTING YOUR BAOFENG: ACCESSORIES THAT MAKE A DIFFERENCE

When you first get your Baofeng radio, it comes ready to use right out of the box, but you can take your experience to the next level with a few key accessories. From antennas that increase your range to batteries and earpieces that improve convenience and usability, Baofeng radios are highly customizable. In this chapter, we'll explore some of the best accessories that can help you get the most out of your radio.

The Best Antennas for Better Range

One of the most effective ways to enhance your Baofeng radio's performance is by upgrading the antenna. The standard antenna that comes with most Baofeng models, often referred to as a "rubber duck" antenna, is serviceable but not ideal for maximizing range. Luckily, switching to a high-performance antenna is an affordable and easy upgrade that can make a big difference in how far your radio signal reaches.

1. **Nagoya NA-771 Antenna**:

   - The **Nagoya NA-771** is one of the most popular aftermarket antennas for Baofeng radios, known for significantly improving signal strength and range. Its longer length compared to the stock antenna allows for better reception, especially in open areas.

   - This antenna works well for both UHF and VHF frequencies, making it versatile for various types of communication, whether you're in the city or out in the wilderness.

   - Installation is simple—just unscrew the original antenna and screw on the Nagoya NA-771. For most users, this antenna upgrade results in a noticeable improvement in signal quality and range, especially when communicating over longer distances.

2. **Abbree Tactical Antenna**:

    ➤ If you need even more range, consider the **Abbree Tactical Antenna**. This antenna is flexible and foldable, making it easy to pack away when not in use, yet strong enough to improve signal strength over greater distances.

    ➤ Its unique design is ideal for outdoor enthusiasts who may need to carry their radio in a backpack or vehicle. The foldable feature also makes it great for users in tight spaces, as you can extend it only when necessary.

    ➤ With the Abbree Tactical Antenna, you'll see improved performance in areas where obstacles like trees and buildings might typically limit your radio's range.

3. **Diamond SRJ77CA**:

    ➤ Another popular option is the **Diamond SRJ77CA**, which is especially well-suited for urban environments. If you're frequently using your Baofeng radio in densely populated areas where interference can be an issue, this antenna can help cut through the noise and provide clearer reception.

- It's also a dual-band antenna, meaning it works with both UHF and VHF, so you won't have to switch antennas depending on the frequency you're using.

- Like the Nagoya, the Diamond SRJ77CA is easy to install and provides a quick and effective way to boost your signal.

By upgrading to a high-quality antenna, you can significantly enhance the performance of your Baofeng radio, ensuring clearer communication and greater range, especially in challenging environments.

Extra Batteries and Earpieces You Should Consider

Having the right accessories isn't just about performance—it's also about convenience. Extra batteries and earpieces can help you stay connected longer and communicate more comfortably in different situations.

1. **Spare Batteries**:

    - If you plan on using your Baofeng radio for extended periods, especially in remote areas where charging might not be an option, investing in spare batteries is essential. The standard battery that comes with most Baofeng models is decent, but it won't last forever, especially if you're using your radio heavily.

- **Baofeng BL-5 Battery Pack**: This is the standard 1800mAh battery that comes with most Baofeng radios, but you can purchase additional packs to ensure you always have power when you need it. With a fully charged spare, you can easily swap out your battery and continue using your radio without interruption.

- **Upgraded Batteries**: For even longer battery life, you can opt for upgraded batteries with higher capacity, such as the **Baofeng BL-5L 3800mAh Battery**. This larger battery provides significantly more usage time, which is especially helpful for longer trips or outdoor activities.

2. **Battery Charging Options**:

- In addition to spare batteries, consider portable charging options. **Battery eliminators** allow you to power your radio directly from a car's 12V socket, which is perfect for road trips or mobile communication. **Solar chargers** are another great option for users who spend a lot of time outdoors and may not have access to traditional power sources.

3. **Earpieces**:

- Earpieces are a game-changer for anyone who wants to use their Baofeng radio discreetly or hands-free. Whether you're in a busy environment or need to keep your communication private, an earpiece ensures you can hear clearly and respond without drawing attention to yourself.

- **Baofeng Earpiece with PTT**: The standard Baofeng earpiece features a **Push-To-Talk (PTT) button** and a built-in microphone, allowing you to easily communicate without holding the radio to your face. This is especially useful in situations where you need to keep your hands free, such as hiking, working, or driving.

- **Surveillance-Style Earpieces**: For more discreet communication, consider a surveillance-style earpiece, similar to what security personnel use. These earpieces fit snugly in your ear and feature a transparent acoustic tube, making them less noticeable while still providing clear audio.

4. **Speaker Mics**:

    - If you prefer not to use an earpiece, a **speaker mic** is another great accessory. This mic clips onto your clothing, allowing you to communicate hands-free while keeping your radio secured to your belt or backpack. It's ideal for situations where you need quick access to your radio without fumbling with the device itself.

Having the right batteries and earpieces ensures you can use your Baofeng radio comfortably for longer periods and in more challenging environments. These accessories add convenience and practicality to your radio experience, making it more versatile for different uses.

Easy Upgrades to Customize Your Radio

One of the things that makes Baofeng radios so appealing is their versatility and the ability to customize them according to your needs. Beyond antennas, batteries, and earpieces, there are a few more simple upgrades that can make your radio even more user-friendly.

1. **Belt Clips and Holsters**:

    - Keeping your radio secure and accessible is important, especially if you're on the move. **Belt clips** and **radio holsters** make it easy to carry your Baofeng without worrying about dropping it or fumbling for it in your bag.

    - Most Baofeng radios come with a basic belt clip, but you can upgrade to more durable and ergonomic options, such as **tactical holsters** that offer additional protection and easy access.

2. **Handheld Speaker Microphone**:

   ➢ A **handheld speaker mic** is an easy way to improve your communication experience. This accessory allows you to keep the radio attached to your belt or backpack while you communicate through the speaker mic. It's particularly useful for those who need to stay mobile but still need access to their radio.

   ➢ Speaker microphones often come with additional features, such as volume control and extra buttons for adjusting radio settings without needing to touch the main unit.

3. **Programming Cable and Software**:

   ➢ If you want to fully customize your Baofeng radio's frequencies and settings, a **programming cable** is an essential tool. This cable connects your radio to your computer, allowing you to use software like **CHIRP** to quickly and easily program channels, frequencies, and other settings.

   ➢ While Baofeng radios can be programmed manually through the keypad, using a programming cable and software is much faster and more efficient, especially if you want to input multiple channels or customize settings for specific needs.

4. **Extended Speaker Mics and Noise-Cancelling Options**:

    ➢ In noisy environments, having a good **noise-cancelling microphone** can make all the difference in communication clarity. Extended speaker mics with noise-cancelling technology help filter out background noise, ensuring your message gets through clearly even in loud settings like construction sites or crowded events.

5. **Waterproof Cases and Covers**:

    ➢ If you're using your radio in harsh outdoor environments, a **waterproof case** or protective cover is a must-have. These cases keep your radio safe from water, dust, and dirt, which can otherwise damage the internal components. Some cases even offer shock resistance, making them ideal for rugged outdoor use.

    ➢ If you're expecting to use your radio in wet conditions, consider a waterproof cover to extend the lifespan of your device.

By investing in these accessories and upgrades, you can tailor your Baofeng radio to suit your specific needs, whether you're a casual user or rely on it for outdoor adventures, emergencies, or work. These simple modifications not only enhance the functionality of your radio but also help it last longer and perform better in various conditions.

Boosting your Baofeng radio's performance doesn't require expensive or complicated equipment. By upgrading your antenna, adding spare batteries, or using an earpiece, you can greatly improve the range, convenience, and versatility of your radio. Small but effective upgrades like belt clips, programming cables, and waterproof cases can make your radio more user-friendly and ready for any situation. With these accessories, you'll have everything you need to make the most out of your Baofeng radio experience.

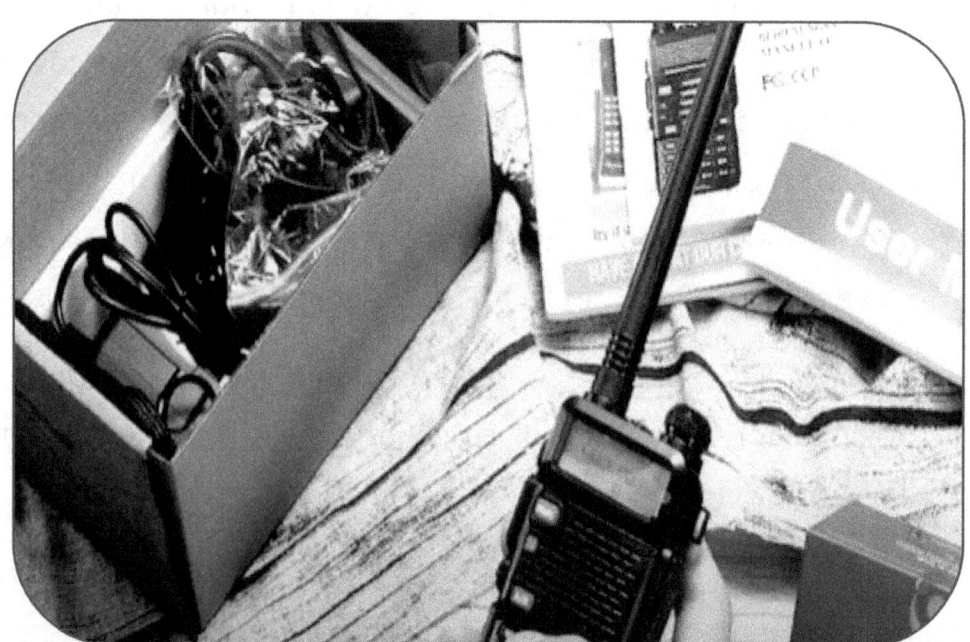

7: IMPORTANT SAFETY AND LEGAL INFO

When it comes to using your Baofeng radio, there are some crucial safety and legal guidelines that you need to understand. Whether you're a beginner or an experienced radio user, following the rules ensures not only that you're operating within the law but also that you're staying safe and responsible in all your communications. In this chapter, we'll cover the basics of whether you need a license, what to know about using your radio in emergencies, and the key safety tips for staying compliant and secure.

Do You Need a License? (Let's Clear This Up!)

One of the most common questions new Baofeng users have is whether or not they need a license to use their radio. The answer is: **it depends on how you plan to use it**.

1. **HAM Radio Frequencies**:

    ➢ If you're planning to use your Baofeng radio on **HAM radio frequencies** (also known as amateur radio frequencies), **yes**, you will need a license. HAM radio operators are required by law to obtain a license from the **Federal Communications Commission (FCC)**

> in the United States or an equivalent regulatory body in other countries.

> ➢ There are three levels of HAM licenses in the U.S.: **Technician, General, and Extra**. For most beginners, the **Technician license** is sufficient to get started, and it covers access to VHF and UHF frequencies, which your Baofeng radio operates on.

> ➢ To get your license, you'll need to pass an exam that covers basic radio theory, regulations, and operating procedures. Fortunately, there are plenty of study guides and resources available to help you prepare.

2. **FRS/GMRS Frequencies**:

> ➢ If you're using your Baofeng radio on **Family Radio Service (FRS)** frequencies, no license is required. These frequencies are designed for short-range, low-power communication and are commonly used for personal and family activities.

> ➢ However, **General Mobile Radio Service (GMRS)** frequencies do require a license, though no exam is necessary. You simply apply for the license through the FCC, pay a fee, and you're good to go.

> ➢ GMRS offers higher power and greater range than FRS, making it a popular choice for outdoor

enthusiasts and families who want more robust communication options.

3. **MURS Frequencies**:

   ➢ **Multi-Use Radio Service (MURS)** is another option that doesn't require a license. MURS operates on a set of five VHF frequencies and is ideal for business or personal use. While not as commonly used as FRS or GMRS, MURS frequencies can be a good choice if you need communication over moderate distances without the need for a license.

4. **Public Safety and Military Frequencies**:

   ➢ It's important to note that Baofeng radios can access frequencies that are **strictly regulated** for public safety, military, and other government communications. **Never use these frequencies** unless you are authorized to do so. Unauthorized use of these frequencies can lead to serious legal consequences, including fines or imprisonment.

In short, while some frequencies are available without a license, if you want to take full advantage of your Baofeng radio's capabilities, obtaining a HAM license is highly recommended. It's a valuable investment that allows you to communicate legally and opens up a world of possibilities in amateur radio.

Using Your Radio in Emergencies: What You Should Know

Your Baofeng radio can be a life-saving tool in emergency situations, but it's essential to understand the rules and best practices for using your radio in such scenarios.

1. **Emergency Frequencies**:

   ➢ In an emergency, you can legally use your radio to **transmit on frequencies you wouldn't normally be authorized to use**, but only if it's a true emergency and no other communication options are available.

   ➢ Some common emergency frequencies include:

   - **146.520 MHz**: This is the national calling frequency for HAM radio operators. It's widely monitored by other HAM users and is a good starting point for seeking help in an emergency.

   - **462.675 MHz**: This GMRS frequency is often used for emergency communication and can be accessed by licensed GMRS users.

- Keep in mind that emergency use should be **brief and to the point**. Clearly state your situation, your location, and the type of help you need.

2. **Preparedness**:

    - If you're using your Baofeng radio in the backcountry or any remote location, it's a good idea to have a few key frequencies pre-programmed for emergencies. This will save you time when you need help quickly.

    - **Have a backup plan**: While your Baofeng radio is a great tool for emergencies, it's always smart to have alternative ways to call for help, such as a satellite phone or personal locator beacon.

3. **Know When to Call for Help**:

    - Use your radio for emergency communications **only when necessary**. In a non-life-threatening situation, it's best to conserve your radio's battery and only call for help when you're sure it's needed.

    - If you're in an emergency situation, stay **calm and clear** when communicating. Avoid panic, and ensure that your message is concise so others can quickly understand your situation.

Staying Safe and Responsible While Using Your Baofeng

Baofeng radios are powerful tools, but with great power comes great responsibility. Staying safe and following the law are key to making sure you get the most out of your radio without getting into trouble.

1. **Respecting Frequency Use**:
   - Always ensure that you're using **authorized frequencies**. If you're unsure whether a frequency is legal for you to use, double-check your country's regulations or consult with a local HAM radio club or online resource.
   - Avoid using **public safety or emergency frequencies** unless you are in an actual emergency. Misuse of these frequencies can block critical communications and put lives at risk.

2. **Battery Safety**:
   - Baofeng radios use rechargeable lithium-ion batteries, which are generally safe but can pose risks if not used properly. **Never overcharge your battery**, and avoid using damaged batteries, as this can lead to overheating or even fire.

- Always carry a spare battery, but make sure it's stored in a way that protects it from extreme temperatures, moisture, or physical damage. A battery pack case or a waterproof bag can help keep your backup power source safe.

3. **Antenna Safety**:

    - When using your radio, be mindful of where you position the antenna. **Never touch the antenna while transmitting**, as this can lead to radio frequency burns or interference with your transmission.

    - Additionally, avoid pointing the antenna directly at your face or body during transmission, as extended exposure to radio waves can pose health risks.

4. **Using Your Radio in Sensitive Areas**:

    - Some locations, such as hospitals or airplanes, restrict the use of radios due to potential interference with sensitive equipment. Always follow posted signs or instructions regarding radio use in these areas.

    - In international travel, be aware that the **legal frequency ranges and rules may differ** from country to country. Make sure you understand the local regulations before using your radio abroad.

5. **Responsible Communication**:

    ➢ Radio communication is meant to be efficient and clear. Avoid using excessive **slang, jokes, or unnecessary chatter** on public channels, especially during peak hours when other users might be trying to communicate.

    ➢ It's also a good practice to monitor a frequency for a moment before transmitting. This ensures you're not interrupting an ongoing conversation or blocking emergency communications.

6. **Know Your Limits**:

    ➢ Baofeng radios are incredibly versatile, but they do have their limits. Understanding the range of your radio and the factors that can impact it—like terrain, weather, and obstructions—will help you make better decisions about when and where to use it.

    ➢ Remember that Baofeng radios are designed primarily for personal or amateur use, and while they are powerful, they should not be relied on as the sole means of communication in critical situations.

Staying safe and legal with your Baofeng radio is crucial to ensuring a smooth and enjoyable experience. Whether you're using your radio for casual communication, in an emergency, or as a HAM radio enthusiast, following these guidelines will help you stay compliant with the law while keeping yourself and others safe.

By obtaining the necessary licenses, understanding emergency protocols, and using your radio responsibly, you can make the most of your Baofeng while avoiding common pitfalls. These radios are fantastic tools for staying connected, but only when used correctly and within the bounds of the law.

8: TOP BAOFENG MODELS FOR 2024

Baofeng radios have earned a solid reputation for offering high-quality, affordable communication devices, but with so many options available, it can be tricky to choose the right one for your needs. In this chapter, we'll go through some of the top Baofeng models for 2024, each with its own set of features, and help you figure out which one is right for you—whether you're a beginner, a seasoned HAM radio enthusiast, or someone who simply needs reliable communication on the go.

Baofeng UV-5R: The Budget-Friendly Choice

The **Baofeng UV-5R** is without trouble one of the most popular models, and for good reason. This radio combines a range of solid features with an unbeatable price point, making it ideal for beginners or casual users who want to get started with HAM radios without breaking the bank.

1. **Affordability**:

    The UV-5R is the go-to choice for those who are looking for a reliable, budget-friendly radio. Priced well under $50, it's perfect if you're not sure how deep you want to get into the HAM radio world just yet.

2. **Dual-Band Capability**:

    Like most Baofeng radios, the UV-5R operates on both

UHF (Ultra High Frequency) and VHF (Very High Frequency) bands. This gives you the flexibility to communicate over different frequencies, ensuring you can reach the people or networks you need.

3. **Compact Design**:
With its lightweight and compact build, the UV-5R is ideal for anyone who needs a radio they can carry with them easily. Whether you're on a hike, at a festival, or just need it for day-to-day use, this model won't weigh you down.

4. **Limitations**:
While the UV-5R is a great entry-level model, it has a relatively lower transmission power compared to more advanced models. It's also somewhat limited in terms of range, making it better suited for short to mid-range communication.

Who is it for?
The Baofeng UV-5R is best for **beginners** or anyone looking for an affordable, easy-to-use radio. It's also perfect for preppers or outdoor enthusiasts who need a basic yet reliable communication tool.

Baofeng UV-82: More Features, Still Simple

The **Baofeng UV-82** is a step up from the UV-5R, offering some extra features while still keeping things simple for users. It's a bit more expensive than the UV-5R but remains one of Baofeng's more affordable models. Here's why it might be a good fit for you.

1. **Enhanced Durability**:
   The UV-82 is designed with a more robust, rugged build compared to the UV-5R, making it ideal for outdoor enthusiasts or anyone who might put their radio through a little more wear and tear.

2. **Improved Audio**:
   One of the standout features of the UV-82 is its **dual push-to-talk (PTT)** button, which allows you to easily switch between two frequencies without needing to reprogram the radio. This can be incredibly handy when communicating with multiple groups or individuals.

3. **Louder Speaker**:
   The UV-82 has an upgraded speaker that produces clearer, louder sound, making it easier to hear transmissions in noisy environments. This is a big plus for those using the radio in the field, at crowded events,

or in situations where communication needs to be crystal clear.

4. **Limitations**:

   While the UV-82 offers more in terms of sound quality and durability, it still has a similar transmission range to the UV-5R. If you're looking for longer range and higher power, you may want to consider a more advanced model like the BF-F8HP.

Who is it for?

The UV-82 is great for those who want a radio that's still simple to operate but with a bit more ruggedness and audio quality. It's perfect for **outdoor adventurers** or **HAM operators** who need a more durable device without sacrificing ease of use.

Baofeng BF-F8HP: Power and Range for Advanced Users

If you need more power and range than what the UV-5R and UV-82 can offer, the **Baofeng BF-F8HP** is your best bet. As Baofeng's high-power option, this model is designed for more advanced users who want the most out of their radio's capabilities.

1. **High Transmission Power**:

   The BF-F8HP offers up to **8 watts of transmission**

power, compared to the standard 5 watts of the UV-5R. This increased power allows for greater range, making the F8HP ideal for users who need to communicate over longer distances.

2. **Improved Battery Life**:
With a larger, 2000mAh battery, the BF-F8HP is designed to last longer between charges. This is especially useful for extended trips or emergency situations where you can't recharge the radio frequently.

3. **Advanced Customization**:
The BF-F8HP is also more customizable than the lower-tier models. You have more options when it comes to programming frequencies and adjusting settings, which makes this model appealing to more advanced HAM users who want complete control over their device.

4. **Limitations**:
The increased power and customization options come with a slightly higher price tag and a steeper learning curve. This model isn't as beginner-friendly as the UV-5R or UV-82, but it's an excellent choice for anyone who has a bit of radio experience under their belt.

Who is it for?
The Baofeng BF-F8HP is designed for **advanced users** who

need more range, power, and flexibility. It's a solid choice for experienced HAM operators, **emergency responders**, or outdoor adventurers who need a reliable, high-power radio for long-range communication.

Easy Comparison: Which One is Right for You?

To help you decide which model best fits your needs, here's a quick comparison of the top three Baofeng models:

Feature	Baofeng UV-5R	Baofeng UV-82	Baofeng BF-F8HP
Price	Low (under $50)	Moderate (under $60)	Enhanced audio
Transmission Power	5 watts	5 watts	8 watts
Range	Short to mid-range	Short to mid-range	Long-range
Durability	Standard	More durable	High durability
Audio Quality	Standard	Enhanced audio	Enhanced audio
Battery Life	Standard	Standard	Longer battery life
Ease of Use	Very easy (beginner-friendly)	Easy (somewhat rugged)	Moderate (more features)

Which One Should You Choose?

- If you're a **beginner** or simply looking for a budget-friendly, easy-to-use radio, the **Baofeng UV-5R** is your go-to. It's perfect for casual users and anyone just starting with HAM radios.

- If you're a bit more **outdoorsy** or need a tougher radio with better sound quality, the **Baofeng UV-82** is a great choice. It's more rugged and offers improved audio for those in louder environments.

- For those who need **extra power** and range, or if you're an **advanced user** looking for more customization options, the **Baofeng BF-F8HP** is the model for you. It offers high transmission power and is ideal for long-range communication.

Baofeng continues to offer some of the best budget radios on the market, making communication accessible to everyone—from beginners to seasoned HAM radio operators. Whether you're looking for a simple, entry-level radio or something with more power and range, Baofeng has a model that fits your needs.

By understanding the differences between the UV-5R, UV-82, and BF-F8HP, you can confidently choose the radio that's right for you and get started with communication that's clear, reliable, and affordable.

9: TAKING CARE OF YOUR BAOFENG

To keep your Baofeng radio working effectively for years, it's essential to take good care of it. While these radios are durable, a little maintenance can go a long way in ensuring optimal performance. In this chapter, we'll explore some simple yet important steps to clean, store, troubleshoot, and maintain your Baofeng radio so that it stays in top shape.

Cleaning and Storing Your Radio (It's easier than you think)

Just like any other electronic device, your Baofeng radio requires regular cleaning and proper storage to function smoothly. Don't worry—this isn't a complicated process, and you don't need any special equipment.

1. **Cleaning the Exterior**

   - **Wipe it down regularly**: Dust, dirt, and grime can accumulate on the surface of your radio, especially if you use it outdoors frequently. Use a **soft, dry cloth** or a slightly damp one to wipe the exterior. Be careful not to let water seep into the buttons or any openings.

   - **Clean the buttons**: Dirt can easily get trapped around the buttons. Use a **cotton swab** or a small brush to gently clean around the edges of each button. If there's any grime buildup, slightly dampen the cotton swab with rubbing alcohol for a more thorough clean.

- **Check the speaker and mic openings**: Dirt or debris stuck in these openings can affect sound quality. Use a brush or compressed air to gently clean these areas.

2. **Storing Your Baofeng Radio**
Proper storage is important to keep your Baofeng radio in good condition when not in use.

- **Store in a cool, dry place**: Moisture and extreme temperatures can damage the internal components of your radio. Always store it in a **dry area**, away from direct sunlight or places with high humidity.

- **Avoid long-term exposure to heat**: Leaving your radio in a hot car or near heating elements can degrade the battery and damage internal circuits.

- **Remove the battery if not in use for a long time**: If you don't plan on using your radio for an extended period, remove the battery to prevent corrosion or battery leakage.

By following these simple cleaning and storage tips, you'll keep your radio looking good and working efficiently.

Troubleshooting Common Issues (Don't panic, I'll Walk you through it)

Even though Baofeng radios are highly reliable, like any piece of technology, they can sometimes experience issues. Don't worry—most common problems can be easily fixed with a little troubleshooting. Below are a few common issues you might encounter and how to solve them.

1. **Problem: Radio Won't Turn On**
   Solution:

   ➢ First, check if the battery is properly inserted. Sometimes, the battery can become dislodged, especially if the radio has been dropped or handled roughly.

   ➢ If the battery is in place but the radio still won't turn on, try charging it. A completely drained battery won't power the radio, even if it was recently used.

   ➢ If the battery won't charge, it may be time to replace it with a new one.

2. **Problem: Weak Transmission Signal**
   Solution:

- Ensure that the antenna is tightly secured to the radio. A loose antenna can reduce the signal strength and range.

- Check your power settings. Some Baofeng models have adjustable power levels (e.g., Low, Medium, and High). Set your radio to **High Power** for stronger transmissions.

- If the issue persists, try changing your location. Obstructions like buildings or hills can block signals. Moving to an open area can improve your range.

3. **Problem: Can't Hear or Be Heard Clearly Solution**:

    - Double-check that you and the other person are on the same frequency and channel. Even being one digit off can result in miscommunication.

    - If the audio sounds garbled, make sure the **Volume** is turned up to an appropriate level.

    - Clean the speaker and microphone as dirt or debris in these areas can block or distort sound.

4. **Problem: Radio Programming Won't Save Solution**:

> This could be an issue with how the programming is being done. Double-check the **steps** you're following to ensure you're saving the frequencies correctly.

> If you're using **CHIRP software**, make sure you have the latest version installed, as older versions may not be compatible with certain Baofeng models.

> Lastly, ensure that the programming cable is working properly, as faulty cables can prevent data from being transferred.

If you run into other issues, check your radio's user manual or look for online forums where other Baofeng users may have experienced and resolved similar problems.

Keeping Your Baofeng in Top Shape

Routine maintenance will ensure that your Baofeng radio continues to perform at its best, even after years of use. Here's how you can keep your radio in tip-top shape.

1. **Battery**   **Care**

   The battery is one of the most critical parts of your Baofeng radio. Here's how to extend its lifespan:

- ✓ **Charge before it's completely drained**: To maximize battery life, try to recharge it when it reaches about 20-30% capacity rather than letting it drain completely.

- ✓ **Use the correct charger**: Always use the charger that came with your Baofeng radio or a compatible replacement. Using the wrong charger can damage the battery or reduce its effectiveness over time.

- ✓ **Avoid overcharging**: It's a good idea to unplug your radio once it's fully charged. Leaving it connected for too long can weaken the battery over time.

2. **Regular Firmware Updates**

   Baofeng radios occasionally receive firmware updates, which can improve performance and fix known bugs. If your model supports updates, make sure you regularly check for new firmware releases. To update:

   - ✓ Use the **CHIRP software** to check for updates.

   - ✓ Follow the instructions in your manual to ensure the update is installed correctly.

3. **Antenna Care**

   The antenna plays a crucial role in your radio's transmission and reception capabilities.

- ✓ **Check for damage**: Occasionally inspect the antenna for signs of wear or damage, especially if you use your radio outdoors frequently.

- ✓ **Upgrade if necessary**: If you're noticing weak signals or limited range, it might be time to upgrade to a higher-quality antenna. A good aftermarket antenna can significantly boost performance.

4. **Environmental Factors**

Take care of your Baofeng radio by being mindful of environmental factors:

- ✓ **Keep it dry**: Although some models are more resistant to water than others, it's always best to keep your radio dry. If it gets wet, turn it off immediately, remove the battery, and let it dry completely before attempting to use it again.

- ✓ **Avoid drops**: While Baofeng radios are sturdy, dropping any electronic device can cause internal damage. Consider using a **belt clip or lanyard** to keep the radio secure and reduce the risk of accidental drops.

Taking care of your Baofeng radio doesn't have to be difficult or time-consuming. By following the simple cleaning, troubleshooting, and maintenance steps outlined in this chapter, you'll keep your radio in excellent condition for years to come. Whether it's storing the radio properly, keeping the battery charged, or troubleshooting common issues, a little attention goes a long way.

A well-maintained Baofeng radio ensures you're always ready to communicate, whether in daily use or during an emergency. So, give your radio the care it needs, and it will serve you well, wherever your adventures take you!

10: FINAL THOUGHTS

As we reach the conclusion of this guide, it's important to reflect on why Baofeng radios are a wise investment and to share some essential tips to maximize your experience with these versatile devices. Whether you're just starting your journey into radio communication or you're an experienced user, Baofeng radios offer a combination of affordability, functionality, and reliability that makes them an excellent choice. Let's explore the key reasons they stand out and review a few final pointers to help you make the most of your radio.

Why Baofeng Radios Are a Smart Investment

Baofeng radios have gained a reputation for being practical and user-friendly, making them suitable for a wide array of users. Here are several reasons why they are a great investment:

1. **Affordability with Quality**

    One of the primary reasons many users choose Baofeng radios is their cost-effectiveness. Compared to other brands, Baofeng radios offer numerous features at a fraction of the price.

    ➢ **Budget-friendly options**: Models like the UV-5R can often be found for under $50, making them accessible for beginners who may not want to invest heavily right away.

> **Great value**: Despite their low price, they are packed with features such as dual-band capabilities and programmable channels, providing excellent functionality for various communication needs.

2. **Versatility Across Different Scenarios** Baofeng radios cater to a diverse range of users, from outdoor enthusiasts to emergency preparedness advocates.

> **Outdoor applications**: Hikers, campers, and hunters appreciate these radios for their portability and durability, especially in areas with unreliable cell service.

> **Emergency readiness**: Preppers find them indispensable, as these radios allow for reliable communication during crises when other forms of communication may fail. Their ability to scan emergency frequencies is a significant advantage.

> **HAM radio community**: For those looking to explore amateur radio, Baofeng offers an affordable entry point into more advanced communication methods.

3. **Durability and Longevity**

Designed with robust materials, Baofeng radios are built to last, making them ideal for frequent use in various environments.

> ➢ **Compact and rugged**: Their small size and sturdy construction make them easy to carry and resilient against wear and tear.

> ➢ **Long-lasting battery**: The batteries generally provide several hours of use on a single charge, ensuring you can rely on your radio in critical moments.

4. **Ease of Use**

Baofeng radios are known for their straightforward operation, making them accessible to individuals with little or no prior experience.

> ➢ **Simple setup**: Most models come with easy-to-follow instructions, allowing you to get started quickly.

> ➢ **User-friendly software**: Programs like CHIRP make it easy to manage and program frequencies from your computer, simplifying the setup process.

In summary, Baofeng radios are a smart investment due to their combination of affordability, functionality, and versatility. Whether you need a radio for casual outings or emergency situations, Baofeng provides a reliable solution.

Last-Minute Tips for Making the Most of Your Radio

Now that you have a solid understanding of your Baofeng radio, here are some valuable tips to help you optimize your experience, ensuring you're prepared for any situation:

1. **Keep Your Battery Charged**
   Always ensure your radio is fully charged before heading out, as a reliable power source is crucial in emergencies or when you're in the field. Consider charging your radio after each use, particularly before outdoor activities.

   ➢ **Extra battery**: Having a spare battery can be invaluable for extended trips or during emergencies.

2. **Regularly Practice Using Your Radio**
   Familiarity is key to confident use. Regularly practice turning your radio on, changing channels, and sending messages to avoid confusion during critical moments.

   ➢ **Emergency frequency knowledge**: Pre-program important emergency frequencies to ensure quick access when needed.

> **Test the range**: Understand the limitations of your radio in your local environment by testing its range and identifying any obstacles that may affect performance.

3. **Keep Your Radio Clean and Safe**
Protect your radio from dirt, moisture, and damage by keeping it clean and stored properly. Regular maintenance will ensure it remains functional for years.

> **Waterproof options**: If you plan to use your radio in wet conditions, consider investing in a waterproof case or cover.

4. **Understand Local Licensing Requirements**
Depending on your location, operating your Baofeng radio may require a license, especially for certain frequencies. Familiarize yourself with local regulations to avoid potential issues.

> **HAM license**: If you're interested in using HAM frequencies, consider studying for a HAM radio license to expand your communication capabilities.

> **Responsible use**: Even if you don't have a license, you can legally use Baofeng radios on designated channels for personal communication and outdoor activities.

5. **Customize Your Radio for Your Needs**

   Make your Baofeng radio your own by customizing it to suit your specific requirements. From programming frequently used channels to upgrading accessories, these small changes can significantly enhance your experience.

   - ➤ **Favorite channels**: Take the time to program channels you frequently use, making access easier in urgent situations.

   - ➤ **Explore accessories**: Consider investing in items like external antennas, earpieces, and carry cases to improve functionality and comfort.

6. **Stay Updated on Firmware and Model Enhancements**

   Baofeng regularly releases updates for their radios, including firmware and new models. Keeping abreast of these updates will help you optimize your radio's performance.

   - ➤ **Follow the community**: Engaging with online forums or user groups can help you stay informed about the latest news and enhancements related to your model.

In conclusion, Baofeng radios are a versatile and affordable solution for anyone interested in enhancing their communication options. Whether you plan to use your radio for emergencies, outdoor adventures, or amateur radio, Baofeng offers various options to meet your needs.

By following the tips and insights shared in this guide, you'll be well-equipped to maximize your radio experience. From basic setup to advanced features, Baofeng radios are user-friendly tools that can improve both your safety and enjoyment of communication. Stay prepared, keep practicing, and relish the many benefits that come with owning a Baofeng radio!

10 interactive questions to your understanding of Baofeng radios so far:

1. What are the two primary frequency bands used by the Baofeng UV-5R?

2. How do you power on the Baofeng UV-5R?

3. What is the purpose of the PTT (Push-to-Talk) button on a Baofeng radio?

4. What does CTCSS stand for, and how is it used in radio communication?

5. What is the role of a repeater in extending your Baofeng radio's range?

6. How do you manually program a frequency into your Baofeng UV-5R?

7. What is the difference between UHF and VHF frequencies?

8. When should you use the squelch feature on your radio, and how does it work?

9. What software is commonly used to program Baofeng radios?

10. Which antenna upgrade can improve your radio's range?

Thank you for taking the time to read the **Baofeng Radio Guide**! I hope this guide has provided you with valuable insights and practical tips for getting the most out of your Baofeng radio. Your decision to invest in this book means a lot, and as a token of my appreciation, I'm excited to offer you exclusive downloadable bonuses to enhance your experience.

To access your **Emergency Communication Plan Template, Glossary of Radio Terms**, and **Emergency Frequency List**, simply scan the QR code below.

Enjoy your new knowledge, and happy communicating!

If you have any questions or suggestions, please feel free to reach me via this QR code too.

www.ingramcontent.com/pod-product-compliance
Lightning Source LLC
Chambersburg PA
CBHW050325230526
45471CB00005B/2363